Pringle, Laurence P
Natural fire: its ecology in forests. Morrow, c1979.
63p. illus. 5.95

Fire ecology
Forest fires
T

NATURAL FIRE

For Lab School students!
Laurence Pringle
2004

NATURAL FIRE

Its Ecology in Forests

by Laurence Pringle

William Morrow and Company
New York · 1979

By the Same Author

Animals and Their Niches
The Minnow Family

Acknowledgment for Photographs

Harold Biswell, 24; National Oceanic and Atmospheric Administration, 41; National Park Service, Bruce M. Kilgore, 46; Packaging Corporation of America, 21 top; Laurence Pringle, 28 top right and bottom, 44, 55; United States Department of Agriculture, 39; USDA Forest Service, 18, 30, 31, 36, 40, Tom Beemers, 28 top left, C. C. Buck, 19, F. E. Dunham, 53, Ray Filloon, 21 bottom, 23, George E. Griffith, 32, Bluford W. Muir, 2, 8, 11, 16, 38, Ed Phillips, 54, H. L. Shantz, 26; USDA Soil Conservation Service, John W. Busch, 49, Al Crouch 13; Wide World Photos, 51

The table on page 50 is adapted from "Fire Management in Grand Teton National Park" by Lloyd L. Loope, *Proceedings of the Annual Tall Timbers Fire Ecology Conference,* Number 14, 1976.

Printed in the United States of America.
1 2 3 4 5 6 7 8 9 10

Library of Congress Cataloging in Publication Data
Pringle, Laurence P
Natural fire.
Bibliography: p. Includes index.
SUMMARY: Explains the beneficial effects of periodic fires to forests and their wildlife.
1. Fire ecology—Juvenile literature. 2. Forest fires—Juvenile literature. [1. Fire ecology. 2. Forest fires. 3. Ecology] I. Title. QH545.F5P74 574.5′264 79-13606
ISBN 0-688-22210-2 ISBN 0-688-32210-7 lib. bdg.

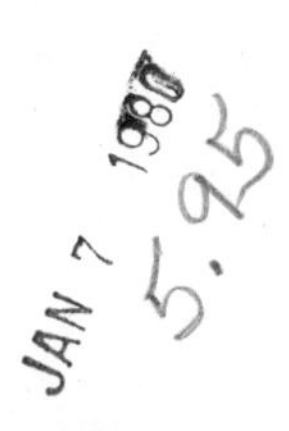

The author wishes to thank Bruce M. Kilgore,
Associate Regional Director,
Resource Management and Planning,
Western Region, National Park Service,
for reading the manuscript of this book
and suggesting changes in it.

CONTENTS

A NATURAL FORCE

A LIGHTNING BOLT flashes in the summer night. It sizzles and spirals down a tree trunk. Wisps of smoke rise from dead pine needles on the forest floor. Flames glow in the night, and a forest fire begins.

The fire spreads quickly. Flames leap up to the crowns of trees, which explode into fireballs. Overhead the fire leaps from tree to tree. A wall of flames moves through the woods, gaining speed. The forest fire seems like a terrible beast with a mind of its own. It roars; it changes direction. It hungrily sucks oxygen from the air and kills almost everything in its path.

Some of the fastest wild animals are able to escape.

The unlucky and the less swift perish—burned to death or robbed of oxygen by the fire. Sometimes a dying rabbit becomes an agent of the fire; its fur ablaze, it dashes crazily through the woods, setting fires as it goes.

At last the fire comes to an end. It dies because of rain, or the efforts of fire fighters, or a combination of factors. But the land is blackened, studded with tree skeletons, littered with dead animals. The soil is vulnerable to terrible erosion, and many years pass before the land heals itself with new plant growth and wildlife.

This scene of death and destruction exists in the imaginations of millions of people—*and seldom anywhere else.* Each year there are more than 100,000 forest fires in the United States. Most are started by people, either accidentally or on purpose. Some are started by lightning. Most lightning-caused fires go out, by themselves, after burning less than a quarter acre of land. And most forest fires of any size are beneficial to plants and animals.

Their good effects have been recognized for many years in the Southeastern United States. Each year forest managers there routinely set ablaze two million acres where pine trees grow. In the West, some wildfires are now allowed to burn for months in national parks and forests. This practice upsets people who feel that all forest fires are "bad."

Just after a fire, this Idaho forest looks lifeless, but many of its plants and animals may have benefited from the blaze.

Whether a forest fire is "bad" or "good" depends on many factors. No one advocates that fires be allowed to burn homes or valuable timber. But fire has been a natural force on land for millions of years, not just in forests but on prairies and savannahs (grasslands mixed with trees and shrubs). Fire became part of our planet's environment as soon as there was vegetation dry enough to be lit by lightning. From then on, periodic fires have been as natural as rain over much of the Earth's land surface. Rain can sometimes be destructive. So can fire. But a great deal of the Earth's plant and animal life has been "born and bred" with fire and thrives under its influence.

After many years of suppressing forest fires at all costs, ideas about them are changing. In the past few decades scientists have learned a lot about ecology—the study of relationships between living things and their environment. This knowledge has led to a new respect for nature and its processes. People are trying to save wild places and to keep them as natural as possible. Some kinds of wild animals have been brought back to places where they once lived, to make these places more ecologically complete. Bighorn sheep, for example, have been reestablished in California's Lava Beds National Monument.

Fire, too, can be allowed back in wild places. But this extraordinary change is not easy to bring about. After

Southern foresters deliberately set surface fires in order to help the growth of slash pine.

many years of fighting fire, the supposed enemy, people must learn how to live with fire, the friend. Instead of putting out all fires, they must learn to recognize and suppress only the harmful ones or harmful parts of them.

People in charge of national parks and other large tracts of land must learn more about the science of ecopyrology—the ecology of fire. And all people need to change their ideas and feelings about forest fires.

THE REAL ROLE

THE STUDY OF FIRE ECOLOGY is complex and fascinating, because there are many kinds of forests and many kinds of fires. To understand the natural role of fire, scientists observe current fires and also investigate fires that occurred centuries ago.

They learn about past fires by examining fire scars. When a fire injures a tree's zone of growing cells (the cambium) and the wound heals, a mark that is eventually covered by bark is left. This scar can be seen later, when the tree is cut down. A cross-section near the tree's base reveals many of the fire scars that formed during the tree's life.

Fires burn bark and leave scars, a record of a tree's fire history, among its annual growth rings.

Studies of these scars show that fire was a normal occurrence in most of the original forests of North America. In California, scientists discovered that fires have happened about every eight years since the year 1685—as far back as they could date the cedar trees studied. They also found that few fire scars had formed after 1900, when people began preventing forest fires.

Ecologists have concluded that low-intensity fires, burning along the ground, were common in Western forests of ponderosa pine and sequoia. They also occurred frequently in Southeastern pine forests. With the exception of swamps and other year-round wet environments, fire used to be a regular happening in many parts of North America.

In many forests of the Pacific Northwest and Northern Rockies, fires were less frequent and usually more intense. Flames reached the crowns of trees, which were often killed or damaged. Forest managers have accepted the idea that such fires are inevitable in parts of the West. The cool, dry climate prevents much decay of dead leaves and other natural litter on the forest floor. Plenty of fuel is available when ideal fire weather occurs, as it does in these northern forests every fifty to one hundred years or so.

Since fires have been a part of forest environments for many thousands of years, many plants are adapted to survive them. These plants might be called "fire species." Among them are the major forest trees of the Northern Rockies: ponderosa pine, white pine, lodgepole pine, larch, and Douglas fir.

These trees have especially thick bark, which can withstand fire damage better than the bark of other

Ceanothus *shrubs thrive where fires occur. Western deer and elk feed on the shrubs, especially in wintertime.*

species. Fire species also include such plants as aspen, willow, and pine grass, which send up many sprouts after suffering fire damage.

One group of Western shrubs, known as *Ceanothus,* is especially dependent on fire. It includes redstem, wedgeleaf, snow brush, and deer brush. *Ceanothus* shrubs are three to nine feet tall and thrive where plenty

Crown fires are inevitable in some Western forests where fuel has accumulated over many years.

of sunlight reaches the forest floor. Once damaged by fire, the shrubs produce abundant new sprouts. Furthermore, *Ceanothus* seeds must be exposed to high temperatures in order to sprout.

Ordinarily, vital moisture cannot get through the hard seed coat to the embryo plant inside. Heat from the fire causes the seed coat to open permanently. The seedling can then develop when conditions are right. After a forest fire, ecologists have counted as many as 242,000 *Ceanothus* seedlings on an acre of land.

The reproduction of jack pine, lodgepole pine, and some other evergreens depends partly on forest fire. These species have sticky resins that hold together the scales of their seed-bearing cones. The cones remain on the trees for many years, storing thousands of pine seeds. In time, a fire releases them. A temperature of about 122 degrees Fahrenheit (50 degrees Celsius) is needed for the resins to melt so that the seeds can pop out onto the ground.

Fire also burns away all or most of the leaves and other natural litter. Many more seedlings grow from such an exposed seedbed than from a surface covered with a deep layer of leaves.

In a plant community that depends on periodic fire, not all species are well-adapted to it. Some take over if fire is kept out of the forest. Without fire, pines in the Southeast are gradually replaced by such deciduous trees as oaks. If no fire occurs for many years in a lodgepole-pine forest in the Rocky Mountains, the old pines are

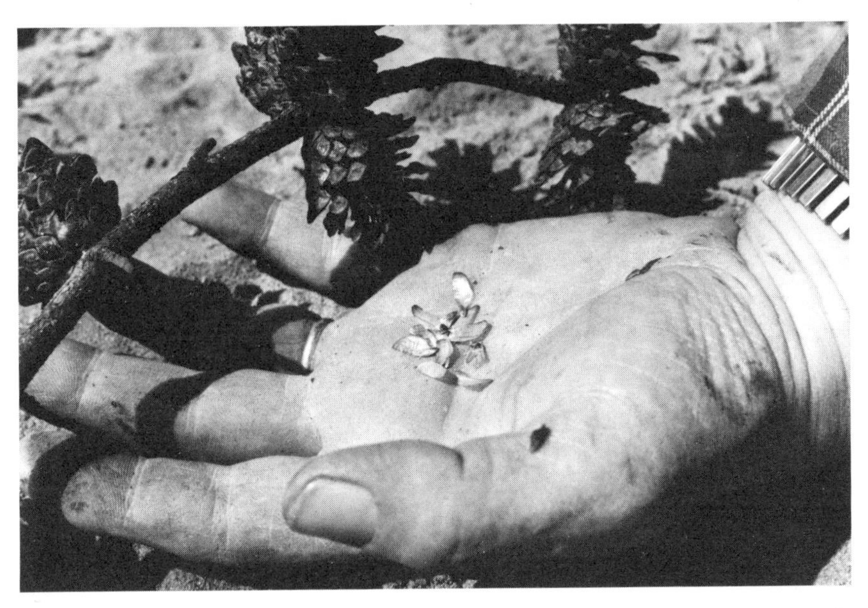

Winged seeds fall from a jack-pine cone that opened as a result of fire. Seedlings of ponderosa pine, below, are abundant on a forest floor where most leaves were burned and seeds fell directly onto soil.

eventually replaced by Engelmann spruce and fir trees. The entire plant community changes unless a forest fire halts the process. A fire would kill many spruce and fir trees, which are less able to withstand the damage than lodgepole pine. And the fire would help release the seeds that represent a new generation of lodgepole pines.

Ponderosa pine is another fire species. It covers thirty-six million acres of Western land, from Nebraska to the Pacific Ocean and from Mexico to Canada. The large needles of ponderosa pine seem designed to encourage fire. Many needles are dropped each year. Because of their size they do not pack down much, and so they dry quickly. They also contain resins. Thus, the needles decay slowly and burn easily.

As long as ponderosa-pine forests have occasional surface fires, the trees thrive and grow in grassy, parklike stands. A ponderosa-pine forest without fire is doomed. When young pines are not thinned out by fire, they grow so close together they are called "dog hair thickets." These thickets are a tremendous crown-fire hazard. Biologists sometimes call these dense stands of trees "biological deserts," because there is so little variety of life in them.

Without fire, white fir and Douglas fir gradually replace the ponderosa pines. The entire forest environment changes. Fir trees have dense crowns, which allow

Influenced by fire, a ponderosa-pine forest has both old and young trees, mixed with grassy openings.

little light to reach the forest floor. Grasses and other surface plants dwindle in numbers and variety—and so do the animals that depend on them. Ecologists have concluded that fire is vital for the survival of beauty and variety in ponderosa-pine forests.

Fire obviously plays a key role in allowing some major plant communities to thrive. Just as there are plant fire species, there are also animal fire species. Elk and deer rely heavily on *Ceanothus* shrubs for winter food in the West. Their health and numbers depend in part on forest fires, which cause *Ceanothus* to thrive. Periodic fires also affect the availability of aspen, a favorite food of moose.

The very survival of the endangered Kirtland's warbler seems to depend on fire. About four hundred of these tiny, colorful birds nest in part of Michigan and nowhere else in the world. They are also known as jack-pine birds, because they build nests under or near young jack pines. This species may never have had a very big range. However, its numbers have declined because of fire control in Michigan jack-pine forests. In an attempt to prevent the Kirtland warblers from dying out, foresters now deliberately plan and set some fires to maintain the kind of nesting habitat needed by them.

Forest fires seldom kill wildlife. Most of them do not occur during the season when birds and other animals

A lack of occasional fires in ponderosa-pine woods may produce a dog hair thicket of young trees.

Elk in Wyoming feed on willow and cottonwood, plants that produce new sprouts after a fire.

have young in nests or dens. Many kinds of animals seem able to sense a fire and its direction, and they move out of its way. Even slow-moving creatures like snakes usually escape.

Every forest fire is different and may have different effects. During most of the fire's life it moves slowly. Rain, lack of wind, or lack of fuel may bring the fire almost to a halt for several days. (Some have been known to smolder all winter long, then resume burning in the spring.) Fires have a daily rhythm too, slowing at night when winds usually die down.

A fire's biography may include a wind-pushed rapid spread when some slow-moving animals are overtaken. It may also burn with great heat in certain areas and suffocate some animals hidden in burrows. Overall, however, wildlife populations are not usually harmed.

Deer, elk, and other large mammals often feed calmly near a surface fire. Usually fire fighters, not flames, are what frighten them away. Foresters working in Southern pine woods report that hawks are attracted by smoke. It may be a signal to them that rodents and other prey are on the move. Eagles and other predatory birds in Africa have also been observed catching insects, lizards, and rodents that are flushed from hiding places by an advancing fire.

Wildlife is attracted to freshly burned land too. Mice and other seed-eating rodents appear in great numbers after a forest fire, sometimes to the dismay of foresters who are concerned about getting a new crop of seedling trees.

For elk, deer, and other plant-eating animals, the end of a forest fire marks the beginning of a period of plentiful and nutritious food. Plants that grow after a fire are usually richer than normal in protein, calcium, phosphate, potash, and other nutrients.

In some ways the burning process is like the process of decay speeded up. As leaves and twigs decay, nutrients are released slowly, over a period of months or years. When leaves and twigs burn, the nutrients are released quickly. From the soil they are gradually recycled into

Deer mice (right) and red squirrels (below) find abundant seeds after a forest fire. These plentiful seed-eating mammals are hunted by hawks (left) and other predators.

the roots of plants. This sudden dose of nutrients shows up in plant tissues for about two years after a forest fire.

Whether the new growth is shrub sprouts, new grasses, or other plants, it is nutritious, tender, and perhaps better-tasting than normal. Elk have been observed eating new sprouts of plants that they usually avoid when the plants are older.

A forest fire also produces a more varied "menu" of plants. The burning away of dead leaves, release of nutrients, and increased sunlight on the forest floor help create an environment in which a great variety of plants can grow. After a forest fire swept through an Idaho, Douglas-fir forest, ninety-nine different kinds of plants appeared where only fifty-one species had been found before.

Ecologists suspect that periodic forest fires have other good effects. Woodsmoke seems to inhibit the growth of fungi, which sometimes harm living trees. Fire also affects populations of insects, including some pests, which spend part of their lives in the leafy litter of forest floors.

Very little is known about this relationship and other aspects of forest-fire ecology. Ecologists wonder how heat affects microorganisms in the soil and the soil itself. Already there is evidence that fires of certain intensities and in certain soil types sometimes cause the formation of a temporary water-repellent layer beneath the soil surface. This layer may cause increased runoff of rainwater and, in turn, soil erosion during the first year after a fire.

This photograph was taken immediately after the Sundance Fire of northern Idaho in 1967.

There are also questions about the nutrients that are released by a forest fire. Are most of them recycled to the forest's plants, or are large amounts washed away? What are a fire's effects on the amount of water in nearby streams and lakes (including reservoirs), on nutrients in the water, and on fish life?

Such questions are under investigation. We already know that an intense fire in a region of heavy rainfall can lead to serious erosion. An example is the Tillamook Burn of Western Oregon, which covered 267,000 acres in

The same place, photographed a year later, is covered by plant growth that was nourished by nutrients from ashes.

1933. Seventeen years after this fire, streams carried five to eight times as much eroded soil as those in similar, unburned areas. Such erosion is unlikely to occur in a region of little rain, however.

Clearly forests and forest fires vary a lot. The climate of a region—even the weather of a particular day—plays a role in the effects of an individual fire. The "behavior" of fires, and how they are affected by fuels, humidity, winds, and terrain, is a fascinating study in itself.

All sorts of knowledge and investigations will be

needed in order to understand and manage the fires that affect the forested one-third of the United States. There is no doubt, however, that the return of periodic fires will be good for most forests and their wildlife.

Intense fire and heavy rain can lead to soil erosion. This picture shows part of the Tillamook Burn in western Oregon.

CHANGING IDEAS

THE JOURNALS OF NORTH AMERICAN EXPLORERS describe the great variety and abundance of plants and animals they found. The continent was not a coast-to-coast forest, but was a patchwork of varied habitats—woods of all ages, mixed with grassy and brushy openings.

The natural variety and beauty they found was partly due to fire. You might say that fire was a native. Lightning-caused fires burned until they stopped. Indians made no attempt to put them out. Indeed, they set fires themselves.

Historians disagree about how often Indians set fires, and exactly why they did so. Like early people everywhere, however, North American Indians saw the good effects of fires and took advantage of them. And they set fires to achieve some of the same benefits. They were the first fire ecologists on this continent.

Colonists from Europe had different ideas. They used fire to help clear forest land for farming, to "tame" wilderness. Unlike the Indians, however, they felt that wildfires were a destructive force and began to protect some lands from them.

This attitude probably contributed to the extinction of the heath hen, a game bird that was once plentiful in the Northeast. It depended on low, brushy habitat (sometimes called heath), which was maintained by periodic wildfires. Without fire, the heath was gradually replaced by forest. The last heath hen died in 1931.

By the early 1900s, public attitudes toward forest fires began to change. Vast forests in the Midwest and West had been cut for timber. Loggers left huge amounts of slash (treetops, limbs, and other leftovers). Once this fuel dried, it was an invitation to disaster. There were a series of devastating fires. In 1871, a fire in Wisconsin covered more than a million acres and took 1,500 lives. The Great Idaho Fire of 1910 burned nearly three million acres and several towns, as well as killing seventy-four fire fighters.

Public alarm over these fires and ruthless logging caused state and Federal governments to begin protecting the land. Our system of national forests and national parks is one result. Government officials also began to

In 1913, this California forest ranger patrolled for fires in a Maxwell runabout.

provide money to hire and train fire fighters, build fire-lookout towers, and buy equipment.

All across the nation, forest fires were declared the enemy. The decision to fight every fire was not unanimous, however. In California, for example, some foresters argued that occasional light fires were needed. They were overruled.

In the motion picture *Bambi* and later in television dramas, forest fires were shown as raging monsters. This belief can be traced back to those huge fires early in this century. It has been reinforced by advertisements that began in the early 1940s. A rugged but friendly-looking bear was chosen as the symbol of the national fire-prevention campaign in 1945. It was named Smokey Bear. The same name was also given to a bear cub that had been found burned and apparently orphaned after a fire in New Mexico in 1950.

This living symbol of forest-fire prevention died in 1976. He had already been replaced at the National Zoo in Washington, D.C., by a younger bear. Smokey's message—"Only you can prevent forest fires"—reaches people from roadside billboards, magazines, and from television. The advertisements often suggest that fawns and other wild animals are frequent victims of forest fires. This appeal to people's feelings has made the advertising campaign very effective. According to the Forest Service, losses from human-caused forest fires have held fairly steady since the early 1960s, which is remarkable because public use of wild lands has been increasing.

Part of this success in preventing forest fires is a result of understanding fires better and having better fire-fighting equipment. Early fire-fighting crews, including smoke jumpers who dropped by parachute near fires in remote areas, had only hand tools.

Today's fire fighters use chain saws and bulldozers to

Dozens of bulldozers may be used to strip the ground of fuel in front of an advancing forest fire.

clear fuel breaks in a fire's path. Mobile radio equipment allows coordination of efforts on the ground and in the air. Helicopters and four-engine cargo planes fly low over a fire and drop chemicals in its path or on hot spots. The largest airplanes can carry 3,000 gallons of fire-retarding chemicals on each trip. The chemicals do not extinguish a fire but slow it until ground forces can arrive.

Helicopters and other modern equipment have made fire control more effective, and much more expensive, too.

Aircraft are also used to detect and observe fires. Planes are equipped with devices that scan the ground below for infrared (heat) radiation. From an altitude of 12,000 feet these scanning devices can detect a blaze the size of a campfire. In addition, they can "see" through the smoke of a large fire and reveal where the heat is most intense. Within minutes, a photographic image of the fire can be sent to fire fighters on the ground.

Infrared film detected the heat of a fire in Topanga Canyon, north of Los Angeles. Knowing the exact location of a fire at a specific time helps efforts to control it.

There has been progress in forecasting fires too. Information about specific forests, the dryness of fuel, and weather conditions can be fed into a computer. It then predicts where fires will most likely occur and how they will "behave" under changing weather conditions.

Lightning is also being studied. Each day about half a million lightning discharges strike forests on earth. Only a small fraction cause fires—about 10,000 forest fires each

"Dry lightning," without rain, is a serious fire threat.

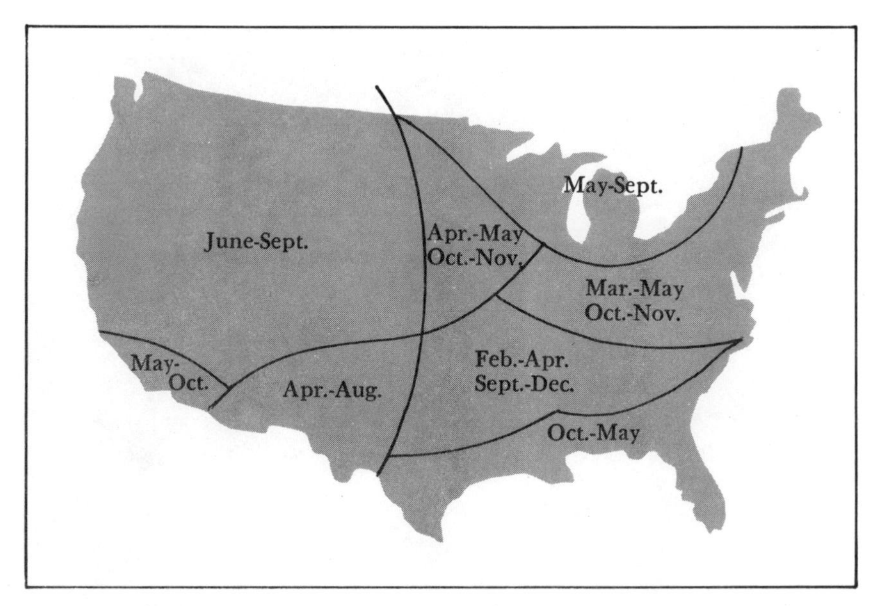

Peak fire seasons vary with the climate.

year in the United States, mostly in the West. In national forests, national parks, and on other public lands in the West, lightning causes about 35 percent of all wildfires. It is the major cause of fire in remote wilderness areas. And these fires usually burn larger acreage and cost more to fight than fires that start near roads.

Scientists now believe that a particular kind of lightning discharge is most likely to start a fire. It must last at least 40 milliseconds. The kind of tree struck is also a factor. Fire is most likely to start when lightning hits a rough-barked evergreen, because the lightning often shatters bark and wood into a shower of tiny bits that becomes easily ignited fuel.

Fire-fighting forces now have electronic devices that

can detect lightning discharges as far as 250 miles away. They can be used to pinpoint the longer-lasting discharges that are most likely to start fires. They can also reveal whether the lightning is accompanied by rain or is "dry lightning," which is the most serious fire threat.

All of these advances certainly help detect and suppress fires. But they have added greatly to fire-control cost too. For example, a 1974 fire covering 470 acres in a national park cost $450,000 to extinguish. During 1978, the Forest Service spent almost $31,000,000 to suppress fires on lands it protects. The annual cost of all forest-fire-control efforts in the United States is well over $300,000,000. In addition, there is a toll of human lives, since nearly all deaths as a result of forest fires are those of fire fighters.

These losses and expenses are sometimes absolutely necessary. Sometimes they are not. Officials who manage forests throughout the United States (and abroad) are reconsidering the effects of trying to extinguish all fires.

Modern fire control is most successful in putting out small fires and light surface fires. During the past few decades, many thousands of such fires, as well as more intense ones, have been extinguished in Western forests. Meanwhile, the fire-dependent plant communities have been doing what they always do—producing plenty of highly flammable fuel. Dead leaves and other fuels have accumulated on forest floors; up to ninety tons an acre have been estimated in the Pacific Northwest and Northern Rockies.

Fallen logs, limbs, and leaves make up most of the fuel that accumulates on the forest floor.

So much fuel is available that there is little possibility of having a light surface fire on millions of acres of forest land. When fires occur they are of unnatural size and intensity. Successful fire control has lead to human-made holocausts. Despite modern fire-control methods and record sums of money spent, forest fires are causing extraordinary damage.

LIVING WITH FOREST FIRES

THE POSSIBILITY OF PROTECTING FORESTS "to death" was foreseen long ago by some foresters. The problem finally received serious attention in the early 1960s, when a scientific report urged a carefully managed return of fire to national parks.

This policy was first attempted in 1968 at Sequoia and Kings Canyon National Parks in California. To allow a return to more natural conditions, 70 percent of these parks—mostly high-elevation wilderness—was designated a "natural fire zone." Lightning-caused fires in this zone are allowed to burn. In other areas of these parks, however, the increase in fuels as a result of past fire suppres-

sion is likely to lead to intense fires. They could threaten survival of fully grown giant sequoias. Here rangers reduce the plentiful fuel by setting carefully planned fires. They avoid causing intense fires by deliberately choosing times of moderate burning conditions.

Nationwide there are more than three million acres of natural fire zones in national parks and monuments. Between 1968 and 1978, some 535 fires were allowed to burn more than 51,000 acres in twelve national parks and monuments. This acreage included not just forest but grassland in Everglades National Park, in Florida, as well. Ecologists and foresters hope that the natural force of fire can also be returned to forests outside of wilderness.

Encouraging fire in forest environments is complicated; much more is involved than simply letting any fire burn, even in an area that has been declared a natural fire zone. Each fire is carefully watched. Parts or all of a fire may be put out if buildings or private lands are in danger.

Managing fire in this way requires a lot of knowledge about forest-fire ecology. The job is not for amateurs. People should still do their best to prevent forest fires. Human-caused fires are still extinguished in national parks, national forests, and other wild lands.

Public attitudes about forest fires are changing, but

Fires are set on purpose to maintain ideal conditions for giant sequoias in national parks.

time is needed for everyone to give up the simple notion that all fires are bad. This problem was shown in the public reaction to the Waterfalls Canyon fire of 1974.

Park rangers allowed this fire to burn for almost three months in Grand Teton National Park. The fire began in early July. In its first month it covered only about 200 acres. By the time it died in early October, it had passed over 3,100 acres. The fire could have been put out easily at any time, which is what some tourists and nearby residents demanded.

Several hundred people signed a petition opposing the "devastation," as they called it. Some of this opposition was a result of the fire's location. Its smoke partly hid the spectacular eastern view of the Grand Tetons, which people expected to see. Other people believed that the fire was destroying timber and wildlife, which was not true. So park officials launched an education program on the natural role of fire in forests.

All across the nation, people are learning that forest fires can be a positive force. The process is slow, however. Reporters still refer to forests "consumed" or "destroyed" by fire, when that description may be far from the truth. The old ideas die slowly.

Getting public support for modern fire management is also difficult because people are worried about air pollution. The controlled burning of Southeastern pine

There is growing controversy over smoke pollution from such controlled burns as this one in a Florida pine forest.

Date	Fire Size	Notes
7/17/74	— —	Last lightning before fire discovered
7/19	— —	Fire discovered
7/20	280 sq. ft.	4 spots covering 1/4 acre
7/22	4,500 sq. ft.	Lightning struck tree toppled causing spot fires
7/24	9,000 sq. ft.	Upslope burning on ground
7/26	6 acres	
7/27	35 acres	Cold front moved in from north
7/28	68 acres	
7/29	80 acres	
7/30	180 acres	Strong north winds
7/31	180 acres	.12 inches precipitation
8/2-8/14	180 acres	A total of .88 inches of precipitation fell on 9 days during this period.
8/18	200 acres	Moderate winds
8/26	230 acres	
9/6	230 acres	Fire moving upslope
9/8	500 acres	High winds, fire ran north 3 1/4 miles
9/11	1,050 acres	Cold front. Very strong winds, first south, then north.
9/11		Snow (.14 inches of precipitation)
9/12		Rain (.13 inches)
9/16	1,650 acres	North wind
9/18	1,900 acres	Fire burned into wind
9/19	2,000 acres	
9/21	2,500 acres	Cold front, north winds
9/22	2,600 acres	
9/26	3,500 acres	Strong southerly winds
9/27	3,500 acres	.15 inches precipitation
10/3	3,500 acres	.28 inches precipitation
10/6	3,700 acres	Date of last spread. A few hot spots into November.

Biography of the 1974 Waterfalls Canyon fire in Wyoming.

forests produces plenty of smoke from December to March each year. So does the burning of slash by loggers in other regions. (It is burned to reduce the hazard of future intense fires.) Forest-fire smoke may lower the quality of the air, which has caused controversy in some states.

The people who may have the greatest problems in learning how to live with forest fires are the residents of southern California. The year 1977 was the second

A house burns in the chaparral-covered hills near Santa Barbara, California.

one of an unusual Western drought. In just two days, lightning started six hundred fires in California.

The most spectacular and damaging fires occurred in southern California. Nearly two hundred homes were destroyed near Santa Barbara. The blaze was the fourth major forest fire near Santa Barbara in thirteen years. In 1978, a fire destroyed about 170 homes near Los Angeles. Such losses seem to be inevitable, as long as people choose to build homes in the hills of southern California.

The plant community that grows on these hills and along the coast of northern Mexico is known as chaparral, or Mediterranean scrub forest. Similar plants grow wherever a climate of mild, wet winters and long, dry summers exists in the world. Most of the shrubs and small trees that make up the chaparral community are fire species. Fire causes them to shed seeds and stimulates the seeds to sprout. Existing plants also produce plentiful sprout growth after a fire. Only a few years are needed for this brushy forest to return to normal after a fire.

Most chaparral plants are highly flammable. Their evergreen leaves contain oils that cause them to ignite and burn easily. After several months without rain—normal weather in southern California—chaparral leaves and stems are very dry. The tiniest spark can start a big, fast-moving fire, especially when hot desert winds are blowing seaward.

California's twenty-four million acres of chaparral are "born to burn" and have had periodic fires for thousands of years. Attempts to prevent fires only postpone it and ensure that the inevitable fire will have more fuel.

Many homes have been built on chaparral-covered hills. Efforts have been made to protect them. Strips of brush-free fuel breaks help keep fires within certain areas. Foresters also burn about 150,000 acres each year in order to reduce the hazard of a later, more damaging fire. These controlled or prescribed burns, as they are called, are also carried out to improve habitat for wildlife and livestock.

A cycle of fire and rapid new growth has always been the natural pattern in chaparral.

Sometimes backfires are set in order to clear fuel in front of a chaparral fire, but they can get out of control, too.

Before starting a controlled burn, foresters take extraordinary care to pick favorable weather conditions. Nevertheless, about one of eight of these fires gets out of control and becomes a chaparral wildfire. There is often a risk that a controlled fire may destroy homes, so this method of avoiding major fires cannot be applied to most places where homes have been built.

Some people seem to accept the idea that they have settled in a plant community of fire species. To keep this problem from worsening, however, further home building may have to be prevented in some areas, which might be designated fire zones. This policy would be like the zoning of flood plains, which has been established in some valleys where damaging floods are inevitable. Chaparral fires are also inevitable.

You can expect to hear further tragic reports of burning homes in southern California. There will be damaging forest fires elsewhere in North America, too, as many years of accumulated fuels explode into flame. But, given the chance, more gentle fires will follow. They will help restore our wild forests to the beauty and variety of long ago.

GLOSSARY

biological desert—an area lacking much variety of plants or animals. The term is usually applied to forests made up almost entirely of trees of the same kind and size.

cambium—cellular tissue in the stems and roots of many plants that produces new cells.

chaparral—dense vegetation of shrubs and small trees with broad, evergreen leaves that grows in such warm, dry regions as southern California, Mexico, and near the Mediterranean Sea.

conifers—trees that bear their seeds in cones made of overlapping scales. Coniferous trees include firs, pines, and spruces.

community—all of the populations of living things occurring in a specific area, for example, in a certain forest or backyard. Ecologists usually specify whether they are referring to a plant community, animal community, or plant-animal community.

crown fire—a forest fire that reaches and travels through the leafy crowns of trees (usually evergreens).

decay—the chemical decomposition of once-living plants and animals.

deciduous—refers to broad-leaved plants that drop their leaves at the end of each growing season, as compared to evergreens, which usually retain their leaves for more than one growing season.

dog-hair thicket—an extraordinarily dense growth of young trees (usually evergreens).

ecology—the study of the relationships between living things and their environment.

erosion—the natural processes by which rocks and soil are worn away and carried away.

evergreens—plants that retain their leaves for more than one growing season. Most conifers are evergreens.

fire species—plants that evolved with periodic fire as a natural force in their environment and that thrive under its effects.

flood plains—low-lying lands along a river channel that are covered with water when the river floods.

habitat—the place where an organism lives, for example, in a young pine-forest habitat or a freshwater pond habitat.

infrared radiation—radiant energy having wavelengths longer than those of visible red light and shorter than those of radio waves. Infrared is invisible to human eyes but can be detected with special films.

nutrients—substances (such as minerals) that nourish and aid the growth, development, and general well-being of living things.

prescribed burns—fires deliberately set under conditions that make them controllable and that are planned to have good effects, for example, to reduce built-up fuels, or to improve habitat for wildlife or certain kinds of desired plants.

predators—animals that kill other animals for food. They include sharks, foxes, robins, ladybugs, and humans.

resins—chemical compounds (such as rosin and amber) in plants. Plant resins are used in varnishes, medicines, and plastics.

savannah—tall grassland, often with scattered trees or patches of woodland, which grows in tropical or subtropical climates.

slash—treetops, limbs, and other leftovers from tree harvesting. Slash is a potential fire hazard and also a potential source of soil nutrients and energy.

FURTHER READING

NOTE: A prime source of information about natural fire is the proceedings of the Tall Timbers Fire Ecology Conferences. Copies can be found in the libraries of some colleges and science museums or ordered from Tall Timbers Research, Inc., Box 160, Tallahassee, Florida 32312.

Anderson, Henry W., "Fire Effects of Water Supply, Floods, and Sedimentation," *Proceedings of the Annual Tall Timbers Fire Ecology Conference*, Number 15, 1976, pp. 249-260.

Biswell, Harold H., "Fire Ecology in Ponderosa Pine-Grassland," *Proceedings of the Annual Tall Timbers Fire Ecology Conference*, Number 12, 1973, pp. 69-96.

Cooper, Robert W., "Status of Prescribed Burning and Air Quality in the South," *Proceedings of the Annual Tall Timbers Fire Ecology Conference*, Number 13, 1974, pp. 309-315.

Fahnestock, George R., "Fires, Fuels, and Flora as Factors in Wilderness Management: The Paseyten Case," *Proceedings of the Annual Tall Timbers Fire Ecology Conference,* Number 15, 1976, pp. 33-69.

Forest Service (U.S.) and National Park Service, *Natural Role of Fire,* 24 pp., 1973. (Available from U.S. Government Printing Office, Washington, D.C. 20402.)

Hall, Frederic C., "Fire and Vegetation in the Blue Mountains: Implications for Land Managers," *Proceedings of the Annual Tall Timbers Fire Ecology Conference,* Number 15, 1976, pp. 155-170.

Handley, Charles O., Jr., "Fire and Mammals," *Proceedings of the Annual Tall Timbers Fire Ecology Conference,* Number 9, 1969, pp. 151-159.

Kilgore, Bruce M., "Fire Management in the National Parks," *Proceedings of the Annual Tall Timbers Fire Ecology Conference,* Number 14, 1976, pp. 45-57.

Komarek, E. V., "The Nature of Lightning Fire," *Proceedings of the Annual Tall Timber Fire Ecology Conference,* Number 7, 1967, pp. 5-41.

Lotan, James E., "Cone Serotiny-Fire Relationships in Lodgepole Pines," *Proceedings of the Annual Tall Timbers Fire Ecology Conference,* Number 14, 1976, pp. 267-278.

Mutch, Robert W., "Wildland Fires and Ecosystems: A Hypothesis," *Ecology,* 1970, pp. 1046-1051.

Nelson, Jack R., "Forest Fire and Big Game in the Pacific Northwest," *Proceedings of the Annual Tall Timbers Fire Ecology Conference,* Number 15, 1976, pp. 85-102.

Taylor, Alan R., "Ecological Aspects of Lightning in Forests," *Proceedings of the Annual Tall Timbers Fire Ecology Conference,* Number 13, 1974, pp. 455-482.

Taylor, D. L., "Some Ecological Implications of Forest Fire Control in Yellowstone National Park, Wyoming," *Ecology,* 1973, pp. 1394-1396.

INDEX

About the Author

A native of Rochester, N.Y., Laurence Pringle attended Cornell University, where he graduated with a B.S. in wildlife conservation. Later, at the University of Massachusetts, he earned an M.S. in the same subject, and he also studied journalism at Syracuse University. Upon finishing his studies, Mr. Pringle taught high-school science for one year and for eight years was editor of *Nature and Science,* a children's magazine published at the American Museum of Natural History in New York City. Since then he has been a free-lance writer, photographer, and editor. At present he lives in West Nyack, New York, in a large house set on an acre of wooded land.